TEACH YOUR KIDS TO COUNT IN ONE DAY!

FUN
ANIMAL NUMBERS

CLAUDIA MOLINA

ONE

THREE

FOUR

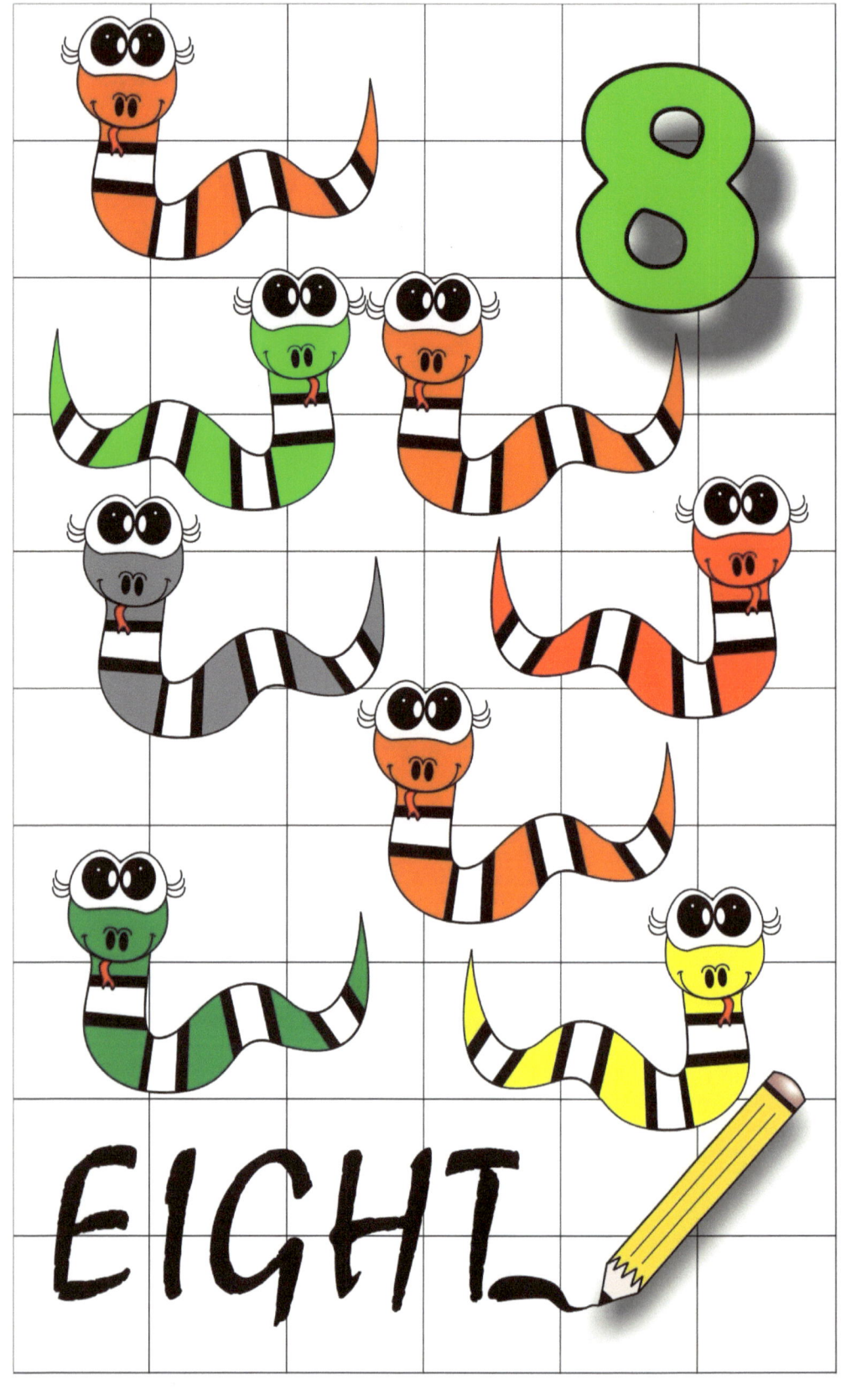

NINE

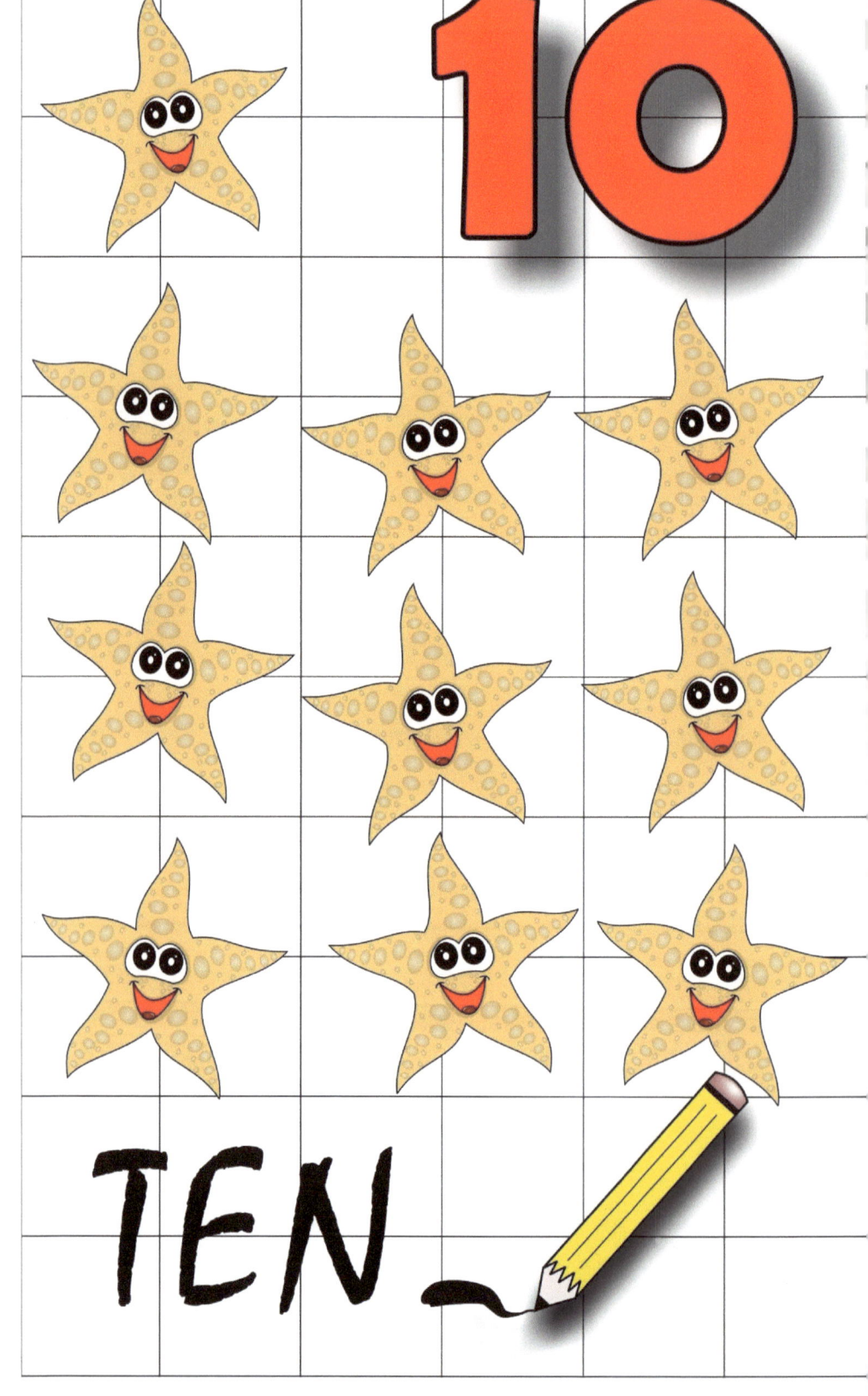

THIRTEEN

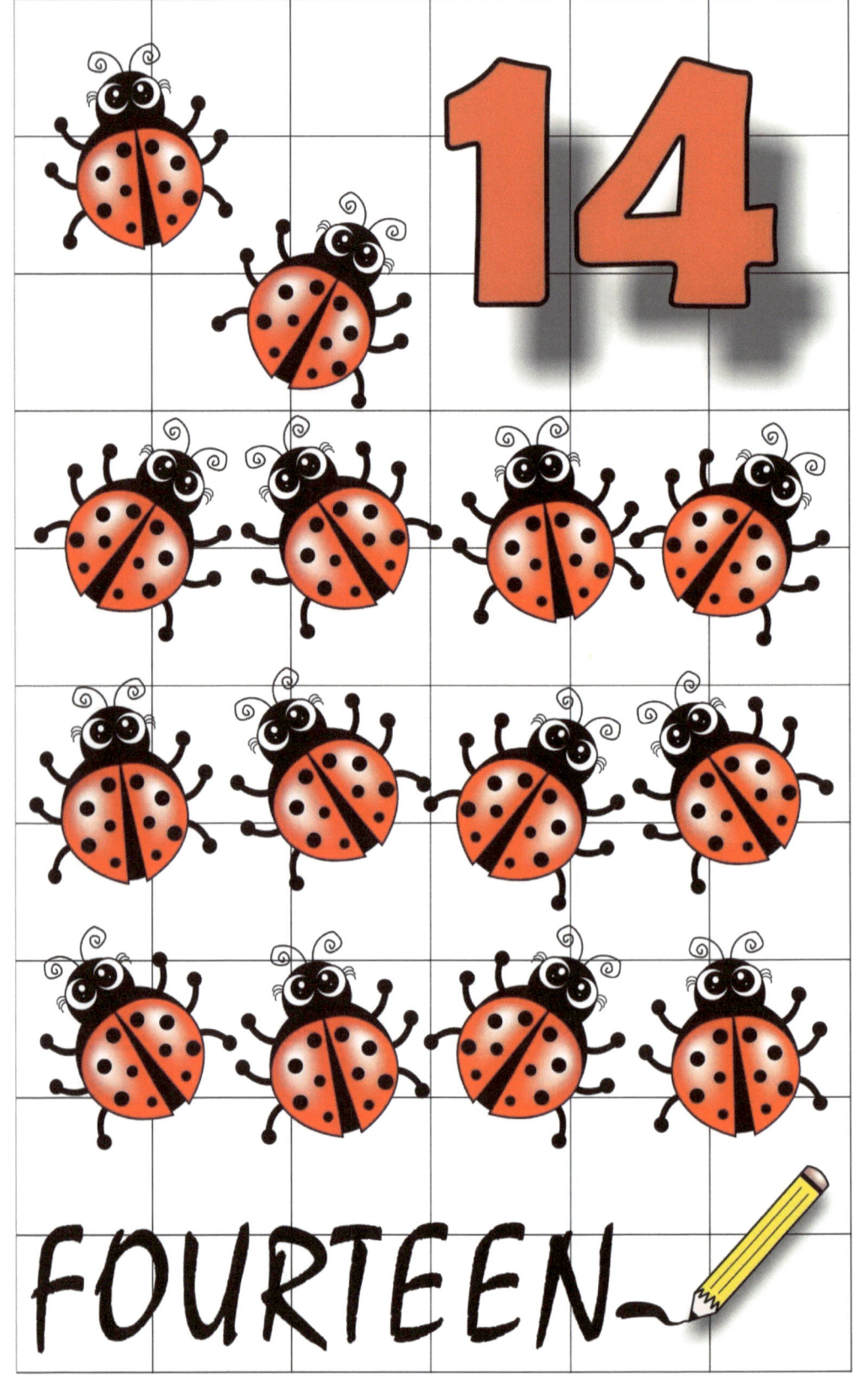

14

FOURTEEN

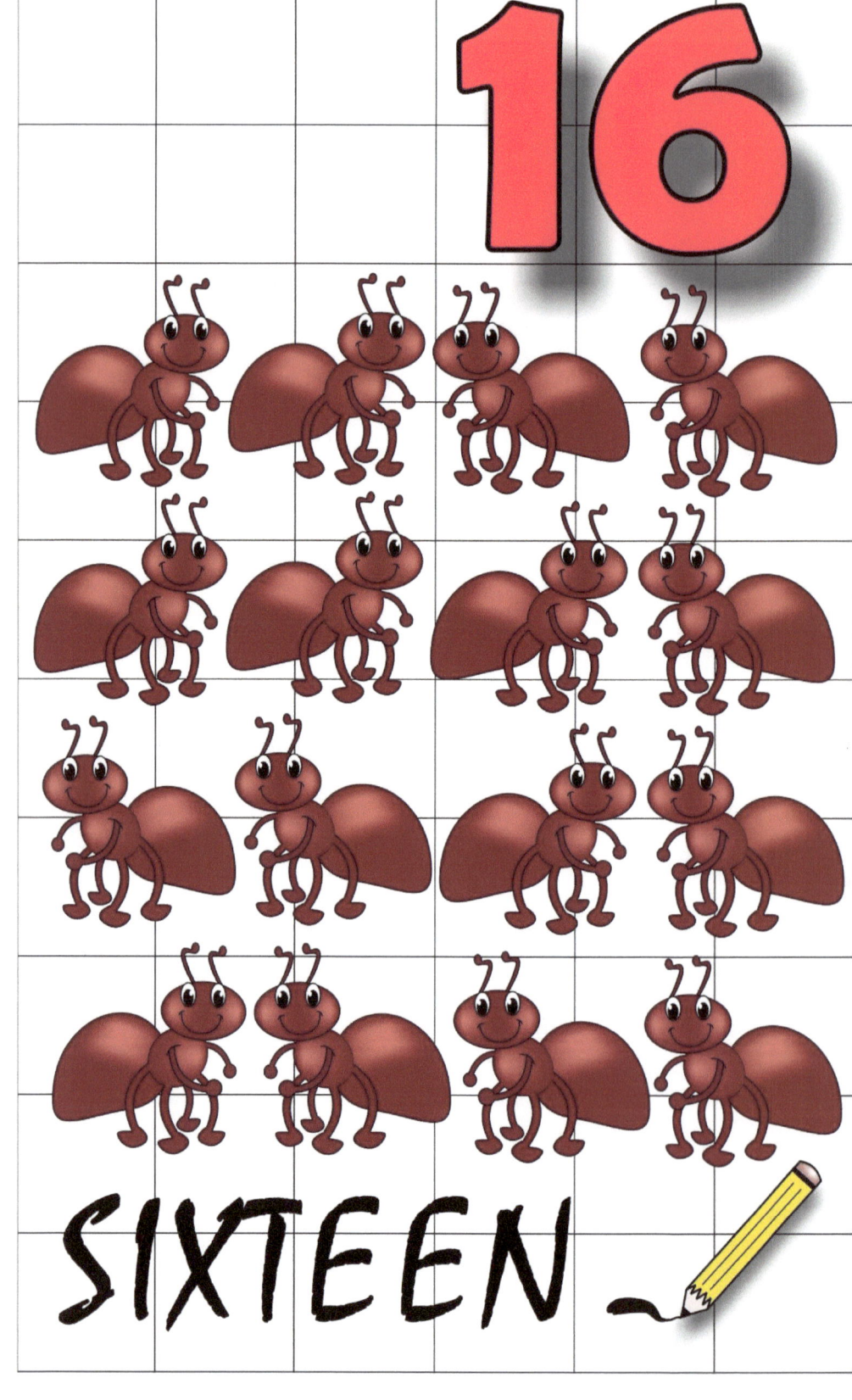

SEVENTEEN

18

EIGHTEEN

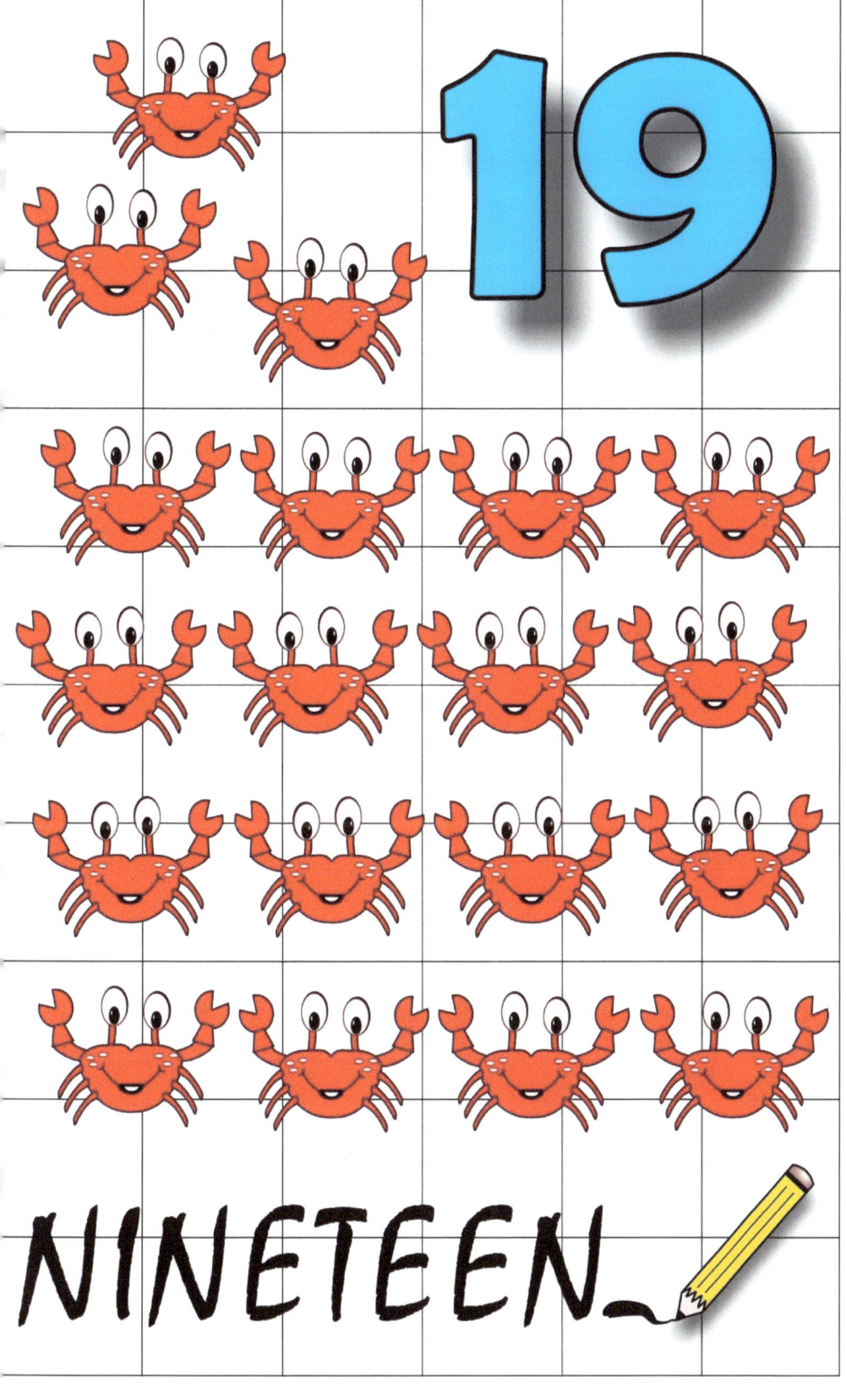

NINETEEN

TWENTY

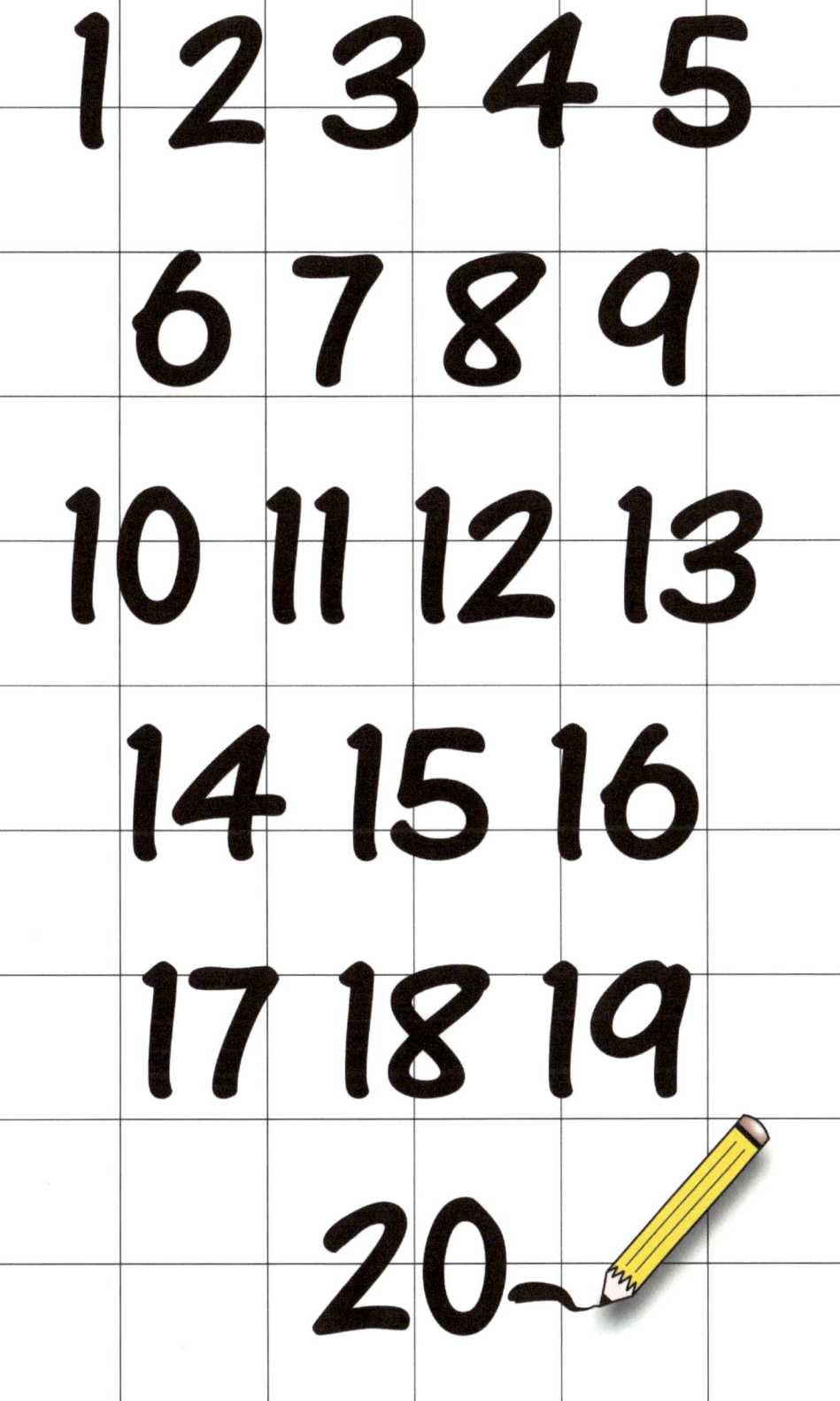

www.ingramcontent.com/pod-product-compliance
Lightning Source LLC
Chambersburg PA
CBHW041618180526
45159CB00002BC/907